The Cherry Fruit-Fly: A New Cherry Pest

Mark Vernon Slingerland

In the interest of creating a more extensive selection of rare historical book reprints, we have chosen to reproduce this title even though it may possibly have occasional imperfections such as missing and blurred pages, missing text, poor pictures, markings, dark backgrounds and other reproduction issues beyond our control. Because this work is culturally important, we have made it available as a part of our commitment to protecting, preserving and promoting the world's literature. Thank you for your understanding.

Bulletin 172. September, 1899.

Cornell University Agricultural Experiment Station,

ITHACA, N. Y.

ENTOMOLOGICAL DIVISION.

The Cherry Fruit-Fly

A

NEW CHERRY PEST

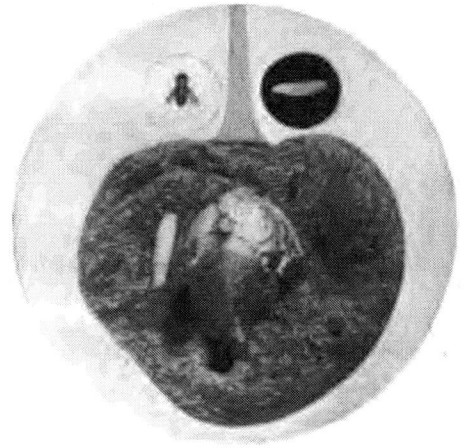

By M. V. SLINGERLAND.

PUBLISHED BY THE UNIVERSITY,
ITHACA, N. Y.
1899.

ORGANIZATION.

BOARD OF CONTROL:
THE TRUSTEES OF THE UNIVERSITY.

THE AGRICULTURAL COLLEGE AND STATION COUNCIL.

JACOB GOULD SCHURMAN, President of the University.
FRANKLIN C. CORNELL, Trustee of the University.
ISAAC. P. ROBERTS, Director of the College and Experiment Station.
EMMONS L. WILLIAMS, Treasurer of the University.
LIBERTY H. BAILEY, Professor of Horticulture.
JOHN H. COMSTOCK, Professor of Entomology.

STATION AND UNIVERSITY EXTENSION STAFF

I. P. ROBERTS, Agriculture.
G. C. CALDWELL, Chemistry.
JAMES LAW, Veterinary Science.
J. H. COMSTOCK, Entomology.
L. H. BAILEY, Horticulture.
H. H. WING, Dairy Husbandry.
G. F. ATKINSON, Botany.
M. V. SLINGERLAND, Entomology.
G. W. CAVANAUGH, Chemistry.
L. A. CLINTON, Agriculture.
W. A. MURRILL, Botany.
J. W. SPENCER, Extension Work.
J. L. STONE, Sugar Beet Investigation.
MARY ROGERS MILLER Nature-Study.
A. L. KNISELY, Chemistry.
C. E. HUNN, Gardening.
A. R. WARD, Dairy Bacteriology.
L. ANDERSON, Dairy Husbandry.

OFFICERS OF THE STATION.

I. P. ROBERTS, Director.
E. L. WILLIAMS, Treasurer.
EDWARD A. BUTLER, Clerk.

Office of the Director, 20 Morrill Hall.
The regular bulletins of the Station are sent free to all who request them.

THE CHERRY FRUIT-FLY.

Rhagoletis cingulata? Loew.

Order DIPTERA; sub-family TRYPETINÆ.

The growing of cherries is already an important phase of the fruit industry of New York and neighboring states. And cherry orchards now frequently supplement the few cherry trees often seen in door-yards, in gardens, or along lanes and roadsides. Everyone who eats this luscious fruit when fresh is familiar with the fact that cherries are often "wormy." Most cherry growers now understand that the cause of "wormy" cherries is that arch enemy of the plum—the plum curculio, shown enlarged in figure 9.

9.—*The plum curculio, enlarged. The insect which "stings" or makes the crescent cut on the cherry, and is responsible for most "wormy" cherries.*

The crescent cut or "sting" of this little beetle is a very discouraging factor to the cherry grower; and the resulting white and footless grub, with a brownish horny head, which revels in the juicy fruit, is a familiar and distracting object to most housewives. In view of these discouraging facts, we are somewhat loath to announce to cherry growers, through the medium of this bulletin, that another, and possibly even a more serious insect enemy, has recently appeared in at least one Massachusetts and in several New York cherry orchards. This new cherry pest works in the fruit, as does the plum curculio, and while it is capable of being equally as destructive, it also works in a much more inconspicuous manner. One can usually readily determine when a cherry is "wormy" from the attacks of the plum curculio, but this new pest gets in its work in such a way that the fruit it infests might easily be classed among the fairest and best on the tree, or in the dish on our breakfast table.

From the above statements, cherry growers can readily understand how serious a menace to their business this new pest might easily become, and how important it will be for them to learn all they can about it.

As we made our first acquaintance with the pest only about two months ago, we have had no opportunity to fully investigate its habits, and hence cannot tell its life-story in detail. For the same reason, we have not tested any remedial measures to control it, but, fortunately, we have at hand the literature giving the results of experiments against similar insect pests working in the fruits of European countries and of our antipodal neighbors in Australia, New Zealand and South Africa. This bulletin is therefore simply a preliminary report for the purpose of calling the attention of cherry growers to this new pest, with an account of what measures have been used against similar pests, all with a view of helping the growers of cherries to understand the nature of the enemy and to be on the lookout for it.

CHARACTERISTICS OF THE NEW PEST.

This new insect enemy of cherries is very different from the plum curculio, which has heretofore been justly accused of being the cause of all "wormy" cherries. The grub of the plum curculio is shown much enlarged in figure 10, while the "worm,"

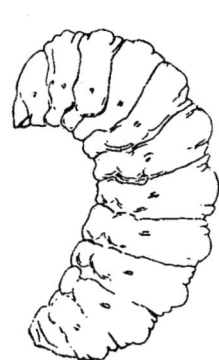

10.—*The grub of the plum curculio, enlarged. This is usually the culprit found in "wormy" cherries.*

which has been found in from one-fourth to one-third of the cherries on some trees the past summer, is shown, natural size and enlarged, in figure 11. As a comparison of these figures will show, this new cherry "worm" is quite different and can be easily distinguished from the *grub* (name applied to the larva of a beetle) of the plum curculio. This new cherry "worm" is instead a true *maggot*, a name given to the larvæ of the two-winged insects—the flies, like the common house-fly.

The shape and size of these cherry maggots, when full-grown, are well shown in figure 11. They are of a very light yellowish-white color. From each side of the body near the head pro-

PLATE I.

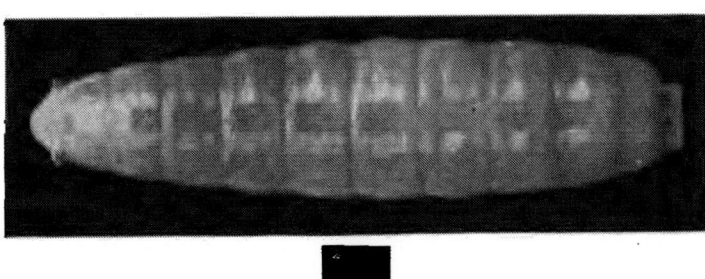

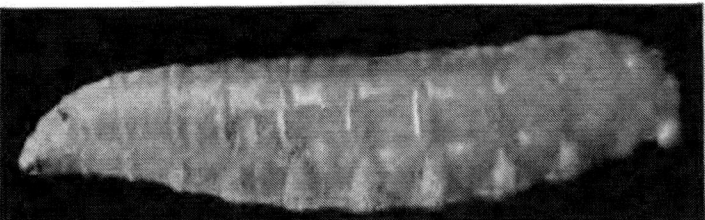

11.—*Dorsal and lateral views of maggot of the cherry-fly. Natural size and much enlarged.*

12.—*Rhagoletis cingulata Leow. The fly which is supposed to be the adult or parent of the cherry maggot. The fly is shown natural size and enlarged, with wings spread and in the normal position when the fly is at rest. The enlarged wing below illustrates a variation in the markings.*

jects a minute, light-brown, fan-shaped organ, which is the cephalic opening of the breathing tubes; the caudal openings of these tubes or trachea form two peculiar, light-brown, slightly elevated, slit-like openings on the caudal end of the body. The mouth-parts consist of two black, minute, sharp, rasping jaws which usually project slightly from the pointed head.

We have as yet found no characteristics by which we can distinguish these cherry maggots from that common pest of the apple—the apple maggot. And we are not yet sure that this new cherry pest is not the apple maggot in a new rolé.

These maggots, which spend practically their whole life in the flesh of the cherry, are the only stage of the insect with which the consumers and most of the growers of the fruit will become familiar.

The maggots hatch from eggs laid by a pretty little fly, resembling in shape, but somewhat smaller than the common house-fly. We cannot know with absolute certainty just what kind of a fly is the parent of the cherry maggot until some of the maggots now in our breeding cages transform into the fly, and this will not take place until next spring. But for reasons to be given later on in discussing the identity of this new pest, we think that the adult form of it is the fly, shown natural size and enlarged, in figure 12. The body of this fly is black, and its head and legs are of a light yellowish-brown color; the lateral borders of the thorax are light yellow; the caudal borders of the segments of the abdomen are whitish; the wings and the scutellum are crossed by four blackish bands and have a blackish spot at their tip; this spot is sometimes confluent with the nearest band, as shown in the enlarged figure of a wing in the lower part of figure 12. The peculiar arrangement of these markings on its wings serves to easily distinguish this fly from any of its near known relatives.

One cherry grower tells us that he saw many of these flies on his trees when the fruit was being picked. He stated that the flies were then somewhat sluggish in their movements, often alighting on the picker's hand. Their black-banded wings render these flies quite conspicuous objects as they flit about from cherry to cherry, so that cherry growers should be able to familiarize themselves with the adult or fly stage of this new enemy.

How and When the Insect Works.

Unfortunately this cherry maggot works in a very inconspicuous manner, so that it will be a difficult matter to determine its presence until the mischief is wrought. All of those who suffered from its ravages the past summer did not know of its presence until their attention was called to it by the con-

13.—*Cherries infested by the Cherry Fruit-fly. All the cherries contained maggots, although the upper ones showed no external indications of being infested. Natural size.*

sumers of the cherries. One grower picked two basketsfull of what seemed to be the fairest and largest cherries, and took them home for canning. When the housewife came to pit them she was much surprised and disgusted to find that many of them were "wormy" with these cherry maggots. The two cherries in the upper part of figure 13 contained maggots, although they were apparently perfect fruits externally. If the cherries are allowed to remain on the tree, or are not used within a few days after picking, the work of the maggot will result in a rotting and sinking in of a portion of the fruit, as is shown by the five cherries in the lower part of figure 13. When this stage is reached, or often even before the fruit shows signs of rotting, the maggots are usually full-grown and soon crawl out of the fruits. One lover of this luscious fruit reports that when some cherries which had been left over from a meal the preceding day, were placed on the table the next morning for breakfast, it was found that several maggots had crawled out during the night. He is now wondering how many maggots were unwittingly eaten the day before!

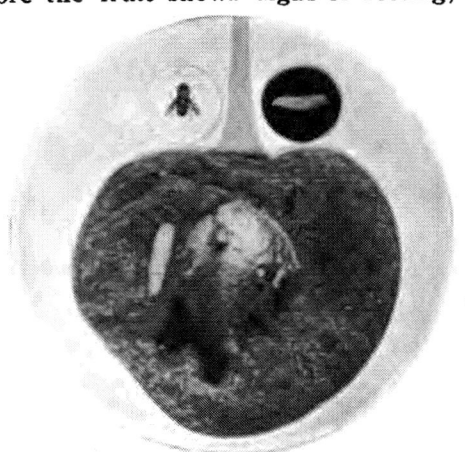

14.—*Section of a cherry, enlarged, to show the maggot and the nature of its work. The small figures above show the maggot and its supposed parent, the fruit-fly. Natural size.*

The work of this cherry maggot is well illustrated in the enlarged picture of a cherry in section, in figure 14. The maggots feed upon the juicy flesh of the ripening cherry, usually near the pit. They form an irregular, rotten-appearing cavity which is represented by the black cavity near the pit in figure 14. Until the maggots get nearly full-grown their work does not show on the surface of the fruit. Soon after "picking-time," however, the rotting extends to the skin which sinks in. Usually but a single maggot is found in a cherry; we have sometimes found a second, but always much smaller, maggot in the same fruit. The maggots do not tunnel all through the flesh of the cherry as does the apple maggot in apples.

We have had no opportunity to ascertain when this cherry maggot begins its work in the fruit. The maggot which works in cherries in Europe is said to begin work about the time the fruits are turning red, and there are indications that our new American pest begins about the same time. It is doubtful if the maggots feed more than three weeks in the fruit, and most of this must be done in the month of June. The maggots may begin their work in the latter part of May in early varieties of cherries, and we have found them in cherries left on the trees as late as August 5th. We also saw many of what we believe to be the adult insect on the cherries at this late date; Mr. Lowe reports finding young maggots in fruits as late as August 16th. Our Massachusetts correspondent reports that some of his cherries began to "spoil" even before they had fully matured.

Varieties of Cherries Attacked.

The European cherry maggot is said to confine its work to the sweet and sub-acid varieties, but its new American congener seems to be less particular in its tastes. The Massachusetts parties who first called our attention to the insect write us that "all our cherries were badly infested, the Downer and the black ones, but the Morellos were the worst." At Ithaca, N. Y., only the early varieties are reported infested; while at Geneva, N. Y., the insect confined its work this year mostly to the English Morello and the Montmorency varieties, the latter being the worst infested. It thus seems that the pest may attack all varieties of cherries whether sweet, sub-acid, or sour, or whether early or late; the Morello and Montmorency varieties seem to have suffered the most this year.

It May Attack Plums or Prunes.

One grower at Geneva, N. Y., reports that he fears the same insect worked in his prunes last year. Ten years ago maggots were found working in both cherries and plums in Northern Michigan. These were thought to have been the apple maggot, but we believe they were identical with those which have worked in the cherries of New York and Massachusetts this year. Our correspondents report that thus far this year they have found no indications of the maggots in their plums or prunes. It would not be surprising to find the maggots working in these fruits, which are oftentimes grown nearby, as they are not very dissimilar in their nature to the cherry. Thus growers of plums and prunes, as well as of cherries, should familiarize themselves with this serious menace to their business. Should anyone find maggots or "worms" of any kind in plums or prunes, we would like to be notified of the fact at once.

Its Distribution and Destructiveness.

We have evidence of the work of this new cherry pest this year from Belmont, Mass., and Ithaca and Geneva, N. Y. The fly which we found on the fruit at Geneva, and which we feel quite sure is the adult insect, is recorded from the Middle States only. It was doubtless the same insect which worked in Northern Michigan ten years ago, as noted above. Thus cherry growers in the Eastern, Middle and Northern States should be on the lookout for the pest.

At Belmont, Mass., about one-third of a six or seven-ton crop of cherries were ruined by the maggots this year. The pest also destroyed from one-fourth to one-third of the crop of English Morello and Montmorency cherries in one orchard at Geneva, N. Y. These facts show that the new pest will become a serious menace to cherry growing in certain sections. Another serious phase of the matter is the fact that the presence of the pest may not be known until the fruit gets into the hands of the consumers, and such fruit will not help in making future sales to the same parties.

Its History, Identity and Name.

So far as we can find there are recorded but two earlier instances where maggots have been found in cherries in America.* For more than a century European cherry growers have suffered from the ravages of a maggot in the fruit. The first record we

---

*Although the bibliography appended to this bulletin includes several references to cherries being found infested by maggots, it may be noted that the records of Cook, Cordley and Davis all refer to the same case of infestation.

find of maggots in cherries in America was made by Dr. Hagen, of Cambridge, Mass., in 1883. That year maggots were very common in the fruit of a black cherry tree imported from Prussia and set in his garden ten years before. He found no differences between his maggots and pupæ and those of the European cherry maggot, but stated that this was not sufficient evidence to prove the specific identity of the two cherry pests. He expected to raise the adult insect and thus settle the identity of our American cherry maggot, but evidently he did not rear the fly, as we are informed that no flies or even any of the maggots are to be found in the collections at Cambridge. It is an interesting fact that we received our first intimation of the existence of such a pest from Belmont, Mass., which is only a few miles from where Dr. Hagen found cherry maggots in 1883.

In 1889, specimens of cherries and plums badly infested with maggots were received at the Michigan Experiment Station from northern Michigan. Brief notices of this infestation were soon published (see bibliography) by Cook, Cordley and Davis. Cordley stated that "from the accounts of our correspondents describing the attack, and from a close examination of both the larva and pupal stages of the insects received, the cherries and plums seem to be badly infested with *Trypeta pomonella* (the apple maggot). Whether these are the descendants of small Trypetas which had formerly acquired a taste for apples, or whether certain individuals of those feeding upon the hawthorn have 'dropped their plebeian tastes and adopted a more refined table regimen,' it is unsafe to say, but from the fact that the apple maggot has never been known to attack the apple of northern Michigan, and from the fact that while the apple maggot is abundant on hawthorn everywhere in Michigan, and as it has not attacked the cherries nor plums elsewhere, it seems probable that a cherry and plum loving race of the apple maggot has developed or is being developed in northern Mchigan, directly from those which fed upon the hawthorn." Unfortunately none of the adult insects seem to have been bred, and we are informed that even none of the maggots are to be found in the Michigan College collection.

Some of the Geneva cherry growers noticed a few maggots

THE CHERRY FRUIT-FLY.

in their cherries last year, and we are informed that they have been seen at Ithaca for a year or more, while our afflicted correspondent at Belmont, Mass., reports that they think their fruit has been infested for the last four or five years, but not nearly so bad as this year.

While there seems to be no evidence extant to ever enable one to determine just what insect is responsible for these two earlier records of maggots in cherries, yet we think the cherry maggots we received this year are the same as those previously recorded. And we furthermore seriously doubt if this new cherry pest is the same as the common apple maggot (*Rhagoletis pomonella*) in spite of the fact that we, like Cordley, have been unable to distinguish between the maggots found in cherries and those working in apple.

The facts recorded by Cordley, as quoted above, strongly indicate that the cherry maggot is a different and distinct insect, and we submit the following evidence in support of this theory. On August 4th we visited an infested orchard at Geneva, N. Y., and found quite a number of English Morello cherries still on the trees, and one or two trees bore many fruits of what the owner called a "sport" or reversion from the English Morello. Many of the fruits contained the maggots, and we soon saw many of the little flies shown in figure 12, on the trees, *almost always on the fruits.* Several of the flies were captured and found to be a species described in 1862 as *Rhagoletis cingulata*, from the Middle States. This fly is thus a very near relative of the apple maggot (*Rhagoletis pomonella*), and a still more significant point is the fact that Loew, in his original description of the fly we found on the cherries, says it is closely allied to the fly of the European cherry maggot. As Doane (1898) has recorded, six species (one, *zephyria*, may prove to be a synonym) of flies of the genus *Rhagoletis* have been described from the United States. The habits are known of only two of these flies, (*R. pomonella*, the apple maggot, and *R. ribicola*, the dark currant fly) and the maggots of these live in fruits.

As we saw no similar flies on the cherry trees, as we found *Rhagoletis cingulata on the fruits*, in considerable numbers, and in view of the facts just submitted regarding the relationships and probable fruit-feeding habits of this fly, it is easy to understand why we have been lead to think that the fly in figure 12 is the adult of our American cherry maggot, and that, therefore, this maggot is a distinct species from the apple maggot. When the adult insects emerge in our breeding cages next spring, our theory, outlined above, may be demolished, as we may get apple maggot flies or something else entirely unexpected, but this will not materially affect the purpose of this bulletin to record all we have been able to glean regarding an insect, whether old or new, which may certainly be classed as a *new cherry pest*.

For this new cherry pest we would propose the popular name of *the Cherry Fruit-fly*. We prefer this name to *the Cherry Maggot* as it is more expressive of the insect's habits, and similar fruit inhabiting maggots in other countries are known as *Fruit-flies*.

POSSIBLE NATURAL FOOD-PLANTS OF THE INSECT.

If this cherry fruit-fly turns out to be the well-known apple-maggot fly, then, of course, its native or original food-plant is the hawthorn. But if this new cherry pest is *Rhagoletis cingulata*, or some insect other than the apple maggot, then we must look to the native species of wild cherries, or possibly wild plums, and also to the species of *Berberis* and *Lonicera* for its natural food-plants. The latter plants are mentioned as possible native food-plants of the American cherry fruit-fly because the European cherry fruit-fly is known to breed in several species of *Berberis* and *Lonicera*.

THE STORY OF ITS LIFE.

Having first made the acquaintance of this new cherry pest only about two months ago, we have had, therefore, no opportunity to follow it through its yearly life-cycle. Hence we are unable to tell the story of its life in detail.

How it spends the winter.—The insect doubtless spends the winter in the soil, usually not more than an inch below the surface, in the condition shown, natural size and much enlarged, in figure 15. It is a dark brown, lifeless-looking object known as a *puparium*. Within this hard, stiff, brown shell, which is really the contracted and hardened skin of the maggot, the insect changes from a maggot to a *pupa*. Whether the pupa is formed before spring, we cannot yet say.

Emergence in the spring.—During the spring months the transformation from a pupa to the adult insect—the pretty little fly shown in figure 12—takes place. When the time for emergence comes, the little fly bursts open one end of the puparium (figure 15,) crawls out, works its way up through the inch or less of soil, and then flits away to find its mate and the food-plant for its progeny. As to when these cherry fruit-flies emerge in the spring we have no evidence. The yellow currant fruit-fly (*Epochra canadensis*) sometimes emerges in May, but the nearer relatives of the cherry fruit-fly, the dark fruit-fly (*Rhagoletis ribicola*) and the apple maggot fly (*Rhagoletis pomonella*) may emerge about the middle of June in the latitude of New York. Hence, we would infer from this that the cherry fruit-fly may be expected to emerge about June 15th, in New York. The date of appearance of the flies on the trees will doubtless vary somewhat with the latitude and the season. The flies will doubtless continue to emerge over a considerable period, perhaps a month or more; the flies which we suspect are the adults of this pest were found on the fruit as late as August 4th.

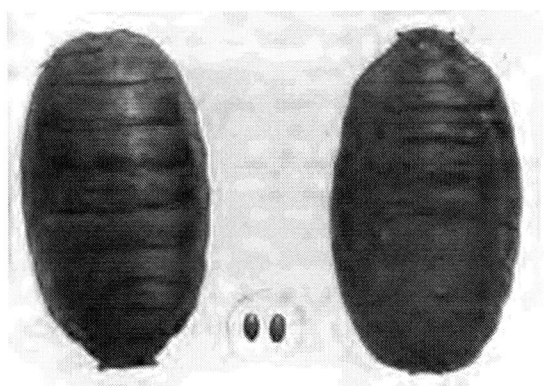

15.—*Dorsal and ventral views of the puparia of the cherry fruit-fly. Natural size and enlarged.*

Egg-laying.—We have not seen the fly lay an egg, but think we have found its eggs in the cherries. We found many minute punctures through the skin of the fruits, and obliquely just beneath the skin in the flesh we could discern the remains of a hatched egg. In a few cases we found an unhatched egg, but always crushed it before we could disengage it from the flesh of the fruit. Hence, we are unable to describe or picture the egg. We feel quite sure, however, that the mother fly punctures the skin of the fruit with her ovipositor and then inserts obliquely an elongate, whitish egg in the flesh just beneath the skin. Mr. Lowe has recorded the following observations regarding the egg : " Egg-laying undoubtedly begins as soon as the first fruit ripens, as young maggots were found in some of the earliest fruits. It continues as late as the middle of August, and probably later. We have found young maggots as late as August 16th. On the same day an unhatched egg was found. The eggs are placed nearly or quite under the skin. One egg was found on the outside. A single egg measured 5 mm. (.02 inch), somewhat broader toward one end, and about one-fourth as wide as long, at the widest point. Beginning at the broad end and extending about one-fourth the length of the egg, the shell is roughened and somewhat darker ; color, a dirty yellow." The flesh of the fruit seems to slightly thicken or harden around the egg and adhere closely to it. Apparently the eggs are laid in any part of the fruit. Old egg-scars are quite easily discernable on the cherries; the minute, round depressed spot on the right-hand cherry of the two upper ones in figure 13 is probably an egg-scar.

Egg-laying doubtless extends over a considerable period, probably beginning in June and continuing until into August, if any cherries remain on the trees so long. We have no data bearing on the duration of the egg-stage. The eggs probably hatch in a few days.

The maggot's life.—As the eggs are laid beneath the skin, the moment it hatches, the maggot finds itself surrounded with its favorite food, the juicy flesh of the fruit. It apparently soon makes its way to near the pit where it proceeds to revel in the flesh, soon forming a rotting cavity, as shown in figure 14. The maggot spends its whole life of three or four weeks in a single cherry, and rarely more than one maggot is to be found in the same fruit. Apparently many of the maggots are nearly full grown about the time the fruit is ready to pick, and they find their way into the consumer's hands. Afflicted orchardists report that but few of the infested cherries fall from the trees, hence when the maggots emerge they doubtless drop to the ground, where they soon bury themselves just beneath the surface. Very soon after entering the ground, probably within a day or two, the maggots contract, their skin hardens and turns brown, and the *puparium* stage is formed. The maggots will change to puparia in any convenient place, as the bottom of baskets, rubbish, etc.

Number of broods.—We have some puparia which were formed in our breeding cages as early as July 11th, from which no flies have yet emerged. Hence we conclude that the insect winters as a puparium, and furthermore, that there is but a single brood of this new cherry pest in a year. Evidently the insect may spend ten or even eleven months of its life in the soil in the *puparia* stage.

How the Insect May be Spread.

As it infests only the fruit, one need have little or no fear of receiving this new cherry pest from nurserymen. If nursery trees happen to be grown under infested cherry trees, it is possible that a few puparia of the pest might be carried away in the soil adhering to the roots of the nursery stock.

As many of the maggots emerge from the fruits, after they reach the consumer's hand, the insect may thus obtain a foothold in new localities. It is quite possible that the insect may be more readily and widely spread in this manner than in any other.

Doubtless the pest will spread quite slowly from tree to tree and thus from orchard to orchard, as the adult insects are slow in their movements and are not long-fliers. This is a very important fact for it makes *the checking of this new cherry pest largely an individual matter, to be worked out independently by each cherry-grower.*

Discussion of Remedial Measures.

It is to be hoped that this new cherry pest is not widely distributed, or that it will never become a serious factor in cherry growing, because it will prove a very difficult pest to control. We have not had time to test any remedial measures, hence can only suggest possible methods, drawn from our experience in combating other insects, from what we know of the habits of the insect, but more especially from the experience of fruit-growers in Australia, South Africa and Europe, where similar fruit-flies are serious drawbacks to fruit-growing.

Apparently there is no possible chance of getting at the insect with a spray of any kind while it is in the egg, in the maggot, or in the puparium stages. The egg is out of reach beneath the skin, in the flesh; the maggot spends practically all its life inside the fruit, only a day or two is spent in getting from the fruit into the soil and changing into a puparium; and this puparium would doubtless be impervious to any liquid applied to the soil in such quantities as not to spoil the soil or injure the tree.

If the maggots caused the infested cherries to fall prematurely, or so affected them as to render it easy to discover which fruits were infested, then one could do much toward controlling the pest by removing such fruits from the trees or by picking up the "windfalls" and destroying them. This latter method can be successfully employed against the apple maggot, which does cause the apples to drop prematurely and which rarely, if ever, leaves the fruit until it does fall or is picked. But afflicted cherry growers state that but few, if any, infested cherries fall prematurely, and also that there is no way of distinguishing the infested cherries from the others at picking time.

Hence there seems to be no practicable method of getting at the pest while it is in the fruit, except the heroic method of picking and destroying by boiling, burying, or otherwise, the whole crop on the infested trees just about the time the first fruits are ready to pick, or even before. This method, of course, involves the loss of the cherry crop for a season, but it is the only sure method we can conceive of to completely check the pest. Usually certain trees or certain varieties will become infested first, and the destruction of the crop on these few trees would not count for much as against their being a constant source of danger to the rest of the orchard. The pest could be quickly stamped out in this way and as it spreads very slowly, it might be a long time in again getting a foothold in the orchard. This method of destroying the crop of cherries for a season, while it is an heroic one, it yet deserves to receive the serious consideration of cherry growers who may be unfortunate enough to have this cherry fruit-fly to combat.

As the insect spends ten months or more of its life in the soil, usually less than an inch below the surface, it would seem as though some method might be devised to check it then. All of this time is spent in the pupariam stage (figure 15), and as we have stated above, while in this form the insect would not be readily affected by insecticides of any sort. We doubt if any of the puparia could be killed by the application of any reasonable or practicable amount of any insecticide, especially such substances as the commercial fertilizers, gas lime, lime or salt. Gas lime has been tried in Australia with no success.

As the puparia are so near the surface of the soil from July until the following June, it would seem as if thorough cultivation might be successfully employed against the pest. But it is evident that the usual methods of cultivation employed by our most successful orchardists has little or no affect on the pest, for those who suffered from it this year were good, thorough cultivators. A possible explanation lies in the fact that the puparia are too small to be crushed, and they are so near the surface that the usual shallow cultivation of the orchard does not materially change their position relative to the surface. Possibly deep plowing, which is not often practicable in a cherry orchard, in late fall or early spring, might bury these puparia so deeply that the emerging flies could not get to the surface. Where only a few trees were infested it would be practicable to remove the surface soil to a depth of an inch or so from beneath the tree and either bury it deeply, put in the hen yard, or in a much-traveled roadway.

One afflicted cherry grower sends the following valuable hint. "We have growing in our hen-yard several cherry trees, and they were not as badly infested as the trees outside of it. We can only account for it in that the hens found the insects as food." Undoubtedly hens would find many of the brown puparia in the soil, and could doubtless be successfully employed against the pest on a few trees. Place a temporary wire-netting fence around one or more trees, turn the hens loose in the enclosure, and stir the soil every day or two to encourage them. Do this soon after the fruit is picked and we doubt if many of the puparia will escape the sharp eyes of the fowls.

Something can be done toward checking the pest by not allowing any cherries to remain on the trees after the last picking. If what few "windfalls" there might be were destroyed, all the marketable fruit picked and disposed of, and all fruits removed from the tree at the last picking, most of the infested cherries would be gotten out of the orchard before most of the maggots had matured and gotten into the soil. Of course, where early and late varieties are infested in the same orchard, this plan might not noticeably diminish the numbers of the pest. It is well worthy of consideration, however.

There yet remains one stage of the insect against which we have not turned our destructive batteries. One of the first questions asked us by an afflicted cherry grower, when he understood how little chance there was of getting at the insect while in the fruit or in the soil, was, why cannot we either kill the flies, or deter or prevent them from laying their eggs in the cherries. He inquired if bad-smelling substances hung in the trees or sprayed upon them would not drive the flies away. Apparently no experiments along this line have been made in this country against similar flies, but, fortunately our Australian and South African fruit-growers, who are sorely afflicted with fruit-flies, have recently carried on valuable and instructive experiments against the flies. The following summary of their experiments and conclusions cannot fail to be of value and interest to our American cherry-growers.

The flies are not attracted to lights in Australia, so that trap-lanterns will be of no avail against the pest.

Messrs. Benson and Voller made a careful and extensive series of experiments in the orchards of Queensland, Australia last year. The objects of their experiments were to prevent or deter the flies from attacking the fruit, and to attract and destroy the flies.

In the first series of experiments they sprayed the fruit and trees with strong smelling substances that were deemed likely to deter or repel the fly. They sprayed with sulphide of lime, sulphide of soda, lime, sulphur, wood tar, bone oil, caustic soda, carbonate of soda, whale-oil soap, tobacco, pyrthum, black leaf tobacco extract, nicotine, and Redwood's specific. Most of the substances were used singly and in various combinations. None of the mixtures injured the fruit or trees to any extent. Many of the mixtures had a very strong and persistent smell, which was retained on the trees and fruit for at least a week after application, and the smell was not washed out by rain, but rather intensified for the time.

Balls of cotton waste saturated with bone oil and other strong smelling substances were also hung in various trees to determine if the odor will deter the flies or not. Flies were seen on fruit within a few inches of the cotton waste, and the trees so treated were as badly infested as any untreated ones.

No spray that was tried was a complete success, even though numerous applications were made; but some mixtures* seemed to keep the flies from the fruit for a certain time after their application, as in the case of the same varieties of fruits, on trees that were sprayed, they were unable to detect a single fly laying eggs, whereas the flies were numerous and busy on adjacent trees. No spray, however, was lasting, as where the applications were made from a week to ten days apart, part of the fruit was infested, but not to the same extent as on untreated trees, thus showing that the applications must be frequent during the ripening of the fruit to be of any avail.

---

*Mixture A :—Boil two pounds of sulphur and one pound of 98 per cent. caustic soda in two gallons of water till the sulphur is dissolved, and a mixture known as sulphide of soda is formed. Add six pounds of whale-oil soap, 80 per cent.; and boil for half an hour, adding boiling water to make five gallons of mixture; and add forty fluid ounces of black leaf tobacco extract. Next add water to make forty gallons, and it is ready for use.

Mixture B :—Dissolve one pound of whale-oil soap, 80 per cent., in four gallons of boiling water. When dissolved, add twenty-five fluid ounces of bone oil and mix well; add water to make forty gallons, and it is ready for use.

Mixture C :—Mix equal parts of A and B.

The experimenters record their belief that careful and frequent sprayings with the mixtures noted above will protect a considerable portion of the crop, but at the same time they are confident that to be of any value the spraying must be very carefully carried out, and must be backed up by destroying all infested fruit and taking every possible precaution to keep the insects in check.

In the second series of experiments made by Messrs. Benson and Voller they tried to attract, catch, or poison the flies. They record that they had no success whatever, as they failed to attract the flies. They used highly-scented sticky baits, highly-scented poisoned baits, and poisoned fruit baits; but, though numerous insects of various kinds were caught or destroyed, the fruit-flies escaped. The experimenters could not find that the flies fed on anything, as, with the exception of seeing them occasionally apparently sucking the juice exuding from a puncture they had made in a fruit, they were never seen to be attracted by or feeding on anything.

In South Africa the only effectual method of preventing the fruit from attacks by these fruit-flies thus far devised is to enclose the trees in a fine-meshed mosquito netting during the time when the flies are about.

We may thus glean from the above summary of the results attained in other countries in combating similar fruit-flies, that there is but little hope of successfully combating our American cherry fruit-fly in the adult or fly state.

No careful experiments seem to have been made in Europe against the European cherry fruit-fly, and the recommendations made for combating the pest are few, usually theoretical, and add nothing new to what we have already suggested, with one exception, which may be of interest to housewives and eaters of the luscious fresh fruit. One German writer states that "it is known to those housewives who wish to can cherries that the maggots leave the fruits as soon as they have been soaked in water for several hours; and this precaution can therefore be taken with the cherries to be eaten fresh in those years when the cherries are badly infested." We wish the author of this suggestion had been a little more definite, for we are in some doubt as to what is to be done with those fruits from which the maggots have emerged. Are they to be canned or eaten with the rest.

MARK VERNON SLINGERLAND

BIBLIOGRAPHY.

1862. Loew. Mon. of Diptera of N. Am., Part I., p. 76. Original description from a female. Habitat, Middle States. States it is closely allied to the European cherry fruit-fly. Figures a wing.

1873. Loew. Mon. of Diptera of N. Am., Part III., p. 263. Briefer description of male and female. Points out great variation in size. States it is closely allied to European *R. flavicincta*. Habitat, Middle States; Long Branch, N. J., in July. Figures a wing.

1878. Osten Sacken. Cat. of Diptera of N. Am., p. 191. References to Loew's descriptions.

1890. Smith. Cat. of New Jersey Insects, p. 398. Quotes Osten Sacken's record of Long Branch, N. J.

1898. Doane. Entomological News, IX., 69-72. Tables for separating the six species of *Rhagoletis*. *R. ribicola* described and compared with *R. cingulata*.

* * * * * *

The following are also included in this bibliography, as we believe they refer to this cherry fruit-fly. They are the only records known to us of the occurrence of fruit-fly maggots in cherries in the American literature.

1883. Hagen. Canadian Entomologist, XV., 159-160. Records *Trypeta* larvæ in fruit of a black cherry tree imported from Prussia; apparently did not differ from those of the cherry fruit-fly (*cerasi*), received from Europe. Did not breed the adults.

1889. Cook. 2d Ann. Rept. Mich. Expt. Station, p. 153. Records receiving plums and cherries from northern Michigan supposedly infested by Apple Maggot.

Cordley. Orchard and Garden, Oct., 1889, p. 192. Records closely examining larvæ and pupæ of plum and cherry maggots from northern Mich. with the result that they seemed to be those of *T. pomonella*. Says *pomonella* has not been known to attack apples in northern Mich., but does occur in haws. Cherry and plums were badly infested.

Davis. *The Ohio Farmer*, Nov. 9, 1889. Records practically same facts as Cordley (1889).

1890. Harvey. Ann. Rept. of Maine Expt. Station for 1889, pp. 192, 233, 234, 235. Records Cook's, Cordley's and Davis' observations and suggests that their plum and cherry maggots may be a distinct species from the Apple Maggot.

1899. Lowe. *Country Gentleman*, LXIV., 693, Aug. 31, 1899. Brief account of the work of the insect at Geneva, N. Y., with description of the different stages.

Slingerland. *Rural New Yorker*, Sept. 16, 1899. Brief, illustrated abstract of this bulletin, No. 172.

Printed by Libri Plureos GmbH in Hamburg,
Germany